How t[illegible] what you want to say ... in mathematics

a guide for students of mathematics who know what they want to say but can't find the words

Lyn Carter and Patricia Hipwell

Count on Numeracy

logonliteracy

First published 2013

National Library of Australia Cataloguing-in-Publication entry

Author:	Carter, Lyn, author.
Title:	How to write what you want to say ... in mathematics / Lyn Carter, Patricia Hipwell ; edited by Catherine Comiskey
ISBN:	9781925046038 (paperback)
Subjects:	Mathematics--Study and teaching. Communication in mathematics.
Other Authors/Contributors:	Hipwell, Patricia, author. Comiskey, Catherine, editor.
Dewey Number:	372.7

Typeset in Delicious 10 pt.

Text and cover design: **Boolarong Press**

Image of *Pencil-pusher* by Zsuzsanna Kilian

Note: Text examples of the writing skill in mathematics have been created to demonstrate that skill. Possible inaccuracies and out-of-date information in these texts are acknowledged by the authors and do not detract from the validity of their inclusion.

Published by Boolarong Press, Salisbury, Brisbane, Australia.

Printed and bound by Watson Ferguson & Company, Salisbury, Brisbane, Australia.

contents

dedication

To our supportive husbands who encourage us to travel the world promoting literacy and numeracy.

introduction

This guide, written by Lyn Carter of **Count on Numeracy** and Patricia Hipwell of **logonliteracy**, provides students with the language they need to write for a range of purposes in mathematics. It aims to provide students with a starting point to say what they want to say using language that mature writers use.

The book is set out in a double-page format:

- **The first page** defines the **writing skill;** provides the **sentence starters** to demonstrate that skill and the **key task words** linked to the skill.
- **The second page** provides words or phrases for **connecting the ideas within and between sentences; useful mathematical language**; an **example** of the **skill** in a short piece of writing.

How to write what you want to say ... in mathematics: a guide for students of mathematics who know what they want to say but can't find the words provides parents, teachers and students with a unique tool for improving writing. It suits students from the middle years of schooling to tertiary level.

This book is the second in a series. The first, *How to write what you want to say: a guide for those students who know what they want to say but can't find the words* by Patricia Hipwell, takes a similar approach to this one but applies to all middle and senior school curriculum areas. A third book, *How to write what you want to say ... in the primary years: a guide for those students who know what they want to say but can't find the words*, is currently in development.

the nature of mathematical writing

The aim of this book is to improve the sophistication of the writing that accompanies or supports mathematical arguments and investigations. There are two important aspects to writing correctly in mathematics: the mathematics must be accurate and the writing must be grammatically correct.

Key features of mathematical writing:

- conveys factual information
- assumes an educated audience
- has a formal, academic tone
- often uses the past tense and passive voice
- avoids personal pronouns, contractions and clichés (see page 42)
- is impartial (avoids bias and opinion)
- includes evidence
- uses mathematical terms such as **mean** and **coefficient**
- includes symbols, tables and visual images
- is a precise and concise form of communication.

key terms and ideas defined

writing skill	the purpose of writing; includes such purposes as describing, comparing, evaluating, justifying
sentence starter	the opening clause of a sentence
key task words	words in questions that establish what is required in the answer (see pages 45-48)
formal language	language that is more characteristic of how we write than how we speak; does not contain colloquialisms or slang; adheres to the conventions of print—grammar, spelling and punctuation are correct
key connectives	words or phrases that link or connect ideas within sentences or ideas from one sentence to the next
modality	expressing ideas such as possibility, certainty, frequency and importance using additional words to extend the main verb
useful mathematical language	some suggested mathematical words and phrases relevant to this skill

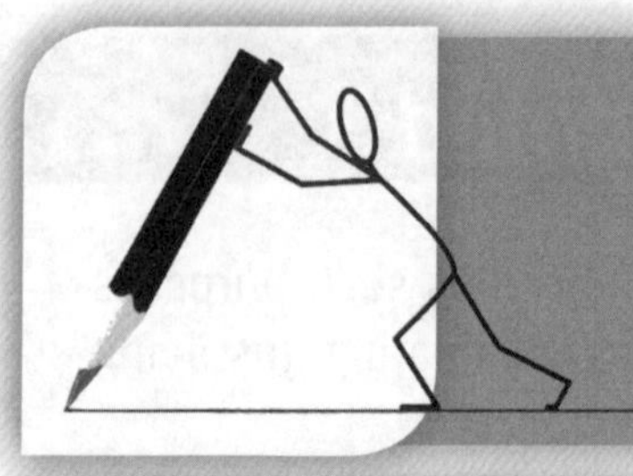

arguing/ persuading

meaning:

presenting one or both sides of an argument and using persuasive techniques to convince others that your opinion about something is the correct one

sentence starters

Nothing polarises opinion as much as the issue of ...

In recent years opinion has become much more divided on the issue of ...

There is a convincing argument/ are several convincing arguments to support this point of view.

Firstly, let us consider the argument that ... — easy to state but difficult to substantiate.

The issue of ... is highly controversial because ...

There has been much debate about ...

The main objection to these alternatives is ...

Given those odds how can there be any doubt?

There is a compelling argument to ...

In spite of/Despite this ...

To argue ... is insufficient, it fails to consider...

Ultimately, it must be realised that ...

All the evidence points conclusively to ...

... has/have been vehemently opposed to ...

There is a great deal of evidence to support ..., not least of which is ...

The evidence that supports this argument is accurate/credible/reliable/unreliable/difficult to substantiate.

It is difficult to decide whether or not ...

Whilst there are several convincing arguments that support this point of view, the balance of the argument is weighted in favour of ...

It would appear that the issue of ... is quite straightforward; however, closer inspection reveals ...

There are compelling arguments both in favour of and against ...

The facts support only one conclusion:

key task words

argue, comment on, construct, debate, exemplify, expound, illustrate, justify, propose, quote, suggest, state

connectives

at one level	because	beyond doubt	even though	evidence suggests
finally	furthermore	hence	however	if ... then
in addition	in conclusion	inevitable/inevitably	moreover	nevertheless
obvious/obviously	of course	on balance	on the other hand	one reason for
so	such as	there are many reasons	therefore	therefore, it seems likely that
ultimately	whilst	without question		

useful mathematical language

absolutely	alternative	agreed/agreement	argued/argument	believed/belief
case	claimed	compelling	conclusive/ conclusively	contradicted
convincing/ convincingly	counter argument	data	debate	demonstrated/ demonstration
disagreed/ disagreement	evidence	facts/factual	limitation	opinion
persuaded	point of view	proved/proof	reasoned/reasoning	refuted
strength	supported	thought	weakness	

example

Nothing polarises opinion as much as the debate about extra-terrestrial life. *However*, **there is a convincing argument** that life must exist elsewhere in the universe. It depends only on what is known about the size of the universe and some very conservative estimates. Scientists generally agree that life, as we know it, requires conditions that are most likely to exist on a planet. An estimate of the number of planets in the universe able to support life is explained in the table below.

	what is known	**conservative assumption**	**conclusion**
galaxies	About fifty billion galaxies are visible by telescope.	There are at least as many galaxies that we cannot see.	There are at least a hundred billion (10^{11}) galaxies.
stars	There are hundreds of billions of stars in an average galaxy.	There are 100 billion stars per galaxy.	There are at least ten sextillion (10^{22}) stars.
planets	Most cosmologists believe that it is common for a star to have planetary systems.	Planetary systems are rare; only one star in one million has a planetary system and each system contains only one planet.	There are at least ten quadrillion (10^{16}) planets.
planets able to support life		It is very rare for a planet to be able to support life: one planet in ten million.	There are at least one billion (10^{9}) planets able to support life.

This table presents **a compelling argument** to show that there could be one billion planets on which life could exist. A billion! To prove the existence of extra-terrestrial life, it need exist on only ONE other planet. **Given those odds, how can there be any doubt that extra-terrestrial life exists?**

classifying

meaning:

grouping so that things with similarities are in the same classes or categories; defending the inclusion of similar things into these categories

sentence starters

Based on ..., the following items can be grouped together like this:

... are/can be classified and named according to ...

This heading/These headings provides/provide a suitable classification for ...

The following classification has been developed for ...

In this group, the following have several attributes in common, and these include ...

This is a very large group, therefore sub-categories can be created.

Common to all of these is the property/characteristic of ...

... can be included in this group because it/they has/have ... in common with the other members of the group.

The following are alike according to the criteria of ...

There are two categories of ... and these are ...

The criterion/criteria for inclusion in this group is/are ...

Despite the superficial characteristic/s of ..., this item merits inclusion in the category of ...

Inclusion in this category is possible because ... has the characteristic of ...

... is the odd one out in this category because ...

The following have the attribute/attributes of ... in common:

The inclusion of ... can be defended by ...

... is classified as ... . On the other hand, ... can be classified as ...

There are clear differences between ... and ..., therefore they belong in different categories.

For the most part, the items in this group have common characteristics that include ...

An alternative classification is ...

There is another way of classifying ...

key task words

arrange, classify, categorise, define, differentiate, discriminate, distinguish, sort

connectives

also	and therefore	and these are	because	but
commonly	except	for the most part	generally	however
if … then	in all cases/instances	in this case/instance	in common	in general
in many cases	in most cases	is based on	is called a	is said to be
on the other hand	so	such as	tend to be	there are examples
typically	usually	whereas	while	whilst
without exception				

useful mathematical language

arranged	associated with	attribute	based on	category/categorised
characterised/ characteristic	class/classification/ classified	criterion/criteria	differed/difference	divided/division
exemplified	grouped/group	hierarchy	included/inclusion	mainly
many	mostly	must	order	organised
property	ranked	set	sorted	sub-divided/sub-division
system	taxonomy			

example

Polygons are two-dimensional closed figures in which all the sides are straight line segments. They **are classified and named according to** the number of sides. Examples are shown in the table below:

number of sides	category
3	triangles
4	quadrilaterals
5	pentagons

number of sides	category
6	hexagons
7	heptagons
8	octagons

Triangles **can be classified according to** the type of angles they contain. *If* all three angles are less than 90°, *then* the triangle **is classified as** acute; *however*, obtuse triangles have one angle exceeding 90° . Right angled triangles must contain a 90° angle. **An alternative classification** *is based on* the side length of the triangle. Equilateral triangles have three sides of equal length, *whereas* isosceles triangles have only two congruent sides. *If* all the sides are different, *then* the triangle *is said to be* scalene.

A classification has also been developed for quadrilaterals. A rhombus has four equal sides, with angles of any size. *However, if* all the angles in the rhombus are right angles, *then* it is called a square. *If* the opposite sides are equal, *then* the quadrilateral **is classified as** a parallelogram, *but* if all of the angles are 90°, the parallelogram is a rectangle. A kite *also* has two pairs of equal sides, *but in this instance* it is the adjacent sides that are equal. *If* one set of opposite sides are parallel, *then* the polygon *is called a* trapezium.

comparing

meaning:

examining two or more things and noting the ways in which they are similar **and** different

sentence starters

There are several ways in which A and B are similar, including ...

The main difference between A and B is ...

The most striking similarity between A and B is ...

Obvious differences exist between A and B, particularly the fact that ...

This differs from ...

Closer inspection reveals that, whilst A and B appear very similar, subtle differences exist.

... and ... have more in common than ... and ..., especially the fact that ...

... and ... are similar as they both show ...

... and ... are different because ... is ...; however, ... is ...

The elements of ... and ... will be compared.

The similarities between A and B are insignificant when compared with their differences.

A comparison of ... and ... reveals noteworthy and highly significant differences.

This is similar to ...

The features of ... and ... are similar, whereas the features of ... and ... are different.

Just as ... is/are ..., so ... is/are ...

In so many ways, A and B are similar, and yet these are often forgotten as the differences are given prominence.

Specific differences exist between ... and ...

The similarities between ... and ... are more relevant than the differences.

They are other reasons why ... are preferred.

key task words

compare, contrast, differentiate, distinguish

connectives

also	alternative/ alternatively	although	as well as	both/in both/all cases
but	compared with/ in comparison	concurrent/ concurrently	despite	even though
however	in contrast	in other respects	in spite of this	in the same way
just as ... so ...	more/greater than	nevertheless	nonetheless	not only ... but also
on the contrary	on the one hand	on the other hand	rather	specifically
the way that	whereas	whilst		

useful mathematical language

alike/unlike	aligned	altered	analogous	assessed
case	changed	comparable	congruent	circumstance
comparison	differed	dissimilar	diverse	equal/equated
event	examined	instance	judged	like
matched	same	similar/similarly	situation	

example

Fractions (called common fractions) **and decimals are similar as they both show** parts of a whole. Fractions use a denominator and numerator to compare the part with the whole. **This differs from** decimals, which use the placement of digits to the right of a decimal point to convey meaning through place value. For example, ¾ is three out of four equal parts, *whereas* the equivalent decimal 0.75 represents seven tenths and five hundredths. The denominator of a fraction may be any value. *In contrast*, a decimal can be written as a fraction only if the denominator is a power of 10 (10, 100, 1000, etc.).

Many students prefer to use decimals as calculators usually display numbers in that form. *However*, without a calculator, fractions are easier to multiply and divide, *whilst* decimals are easier to add and subtract. Some decimals, called recurring decimals, cannot show a value exactly, *whereas* fractions can always show an exact value (for example, the decimal 0.33333... and the fractional alternative of ⅓).

There are other reasons why decimals are often preferred. The decimal form of a mixed number is simpler to use. They do not require reduction to the lowest form. Decimals are also are easier to use with money, percentages and metric system measurements. However, *despite* the preference for decimals, students must be adept in the use of both types of numbers.

contrasting

meaning:

examining two or more things and focussing on the differences

sentence starters

There is nothing about ... and ... that is in any way similar.

The main differences between ... and ... are ...

.... and ... have nothing in common; only differences are apparent.

Obvious differences exist between A and B, particularly the fact that ...

This differs from ...

Differences exist in ...

... and ... are different because ... is ..., whereas ... is ...

Specific differences exist between ... and ...

The elements of ... and ... are very different.

In no way similar is ...

Subtle differences exist between ... and ...

Whilst ... is like ..., ... is like ...

... and ... have far less in common than would at first appear.

The distinguishing characteristics of ... make it very different from anything else.

Do not overlook the hidden differences between ... and ...

... and ... are not alike in any way.

A comparison of ... and ... reveals only differences.

... is at variance with ...

Despite these clear differences ...

... is dissimilar from ...

... is not like ... in any way.

key task words

contrast, differentiate, distinguish

connectives

alternatively	although	conversely	despite	even so
even though	far from	however	in contrast	in fact
in no way similar	in other respects	in spite of this	more/greater than	nevertheless
no commonality	not only ... but also	on the contrary	on the one hand	on the other hand
opposing	or	other differences	rather	whereas
while	whilst	yet		

useful mathematical language

altered	assessed	case	changed	compared with/in comparison
circumstance	contrasted/ in contrast	differed/different/ difference	dissimilar	disparate
distinct	distinguish	diverse	event	examined
instance	judged	nothing like	situation	unequal
unlike	unrelated			

example

There is nothing about perimeter and area that is in any way similar. *On the one hand,* perimeter is the distance around the boundary of a plane shape. *On the other hand,* area is the space within the boundaries of that shape.

Differences exist in the ways that perimeter and area are calculated. This is best illustrated in the case of a rectangle. The perimeter is found by ADDING the lengths of the four sides (although it may be written as $P = 2l + 2w$ or $P = 2(l + w)$ because the opposite sides are equal). *In no way similar is* the method of calculating area that relies on MULTIPLYING the lengths of the different sides ($A = lw$).

Perimeter is a length, and is measured using units of length such as millimetres, centimetres, metres and kilometres. *In contrast,* area is measured using units such as square centimetres, square metres, hectares and square kilometres.

Despite these clear differences between perimeter and area, the two ideas are often confused.

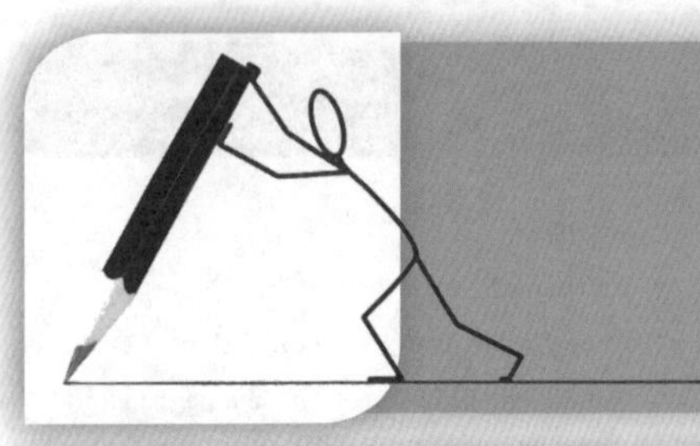

defining

meaning:

to give the meaning or precise description of the concept

sentence starters

The definition of (thing being defined) is ...

A (thing being defined) is made up of ...

A (thing being defined) comprises ...

(thing being defined) can be defined as ...

(thing being defined) is said to be

A (thing being defined) is something that ...

(thing being defined) occurs when ...

A (thing being defined) is a number that can have ...

(thing being defined) is an amount that ...

In (situation), the (thing being defined) is ...

The abbreviation ... means ...

The meaning of (thing being defined) is ...

The symbol ... means ...

... is represented by the symbol ...

... is abbreviated as ...

The name given to ... is (thing being defined)

Another word for ... is (thing being defined)

Values of ... that are not ... are said to be (thing being defined)

An example of (thing being defined) is ...

... is an example of (thing being defined)

key task words

define, explain, describe, state

connectives

excludes	for example	for instance	if	if and only if
is	is called	is not	makes	neither … nor
only	other than	shows	so	such that
that	that can be	that can only be	that has	where
whereas	when	with		

useful mathematical language

characteristic	converse/conversely	defined as/definition	described	designated
example	identified	includes	law	means/meaning
object	observed	process	property	set
something	term	type of	uses	

example

Definition in common language:
A prime number is **the name given to** a counting number, *other than* 1, *that can only be* divided evenly by 1 or itself. A composite number *is a* counting number *that can be* divided evenly by numbers *other than* 1 and itself. The number 1 is *neither* prime *nor* composite.

Definition in mathematics language:
On the set of natural numbers, N, a prime number, $p > 1$, has no divisors *other than* 1 and p itself. **Values of** $p > 1$ **that are not** prime **are** composite.

Definition by use of an example:
An example of a prime number **is** 7. The *only* divisors of 7 are 1 and 7, making 7 a prime number, *whereas* the number 12 has divisors 1, 2, 3, 4, 6 and 12, so it *is not a* prime number. The number 12 **is an example of** a composite number.

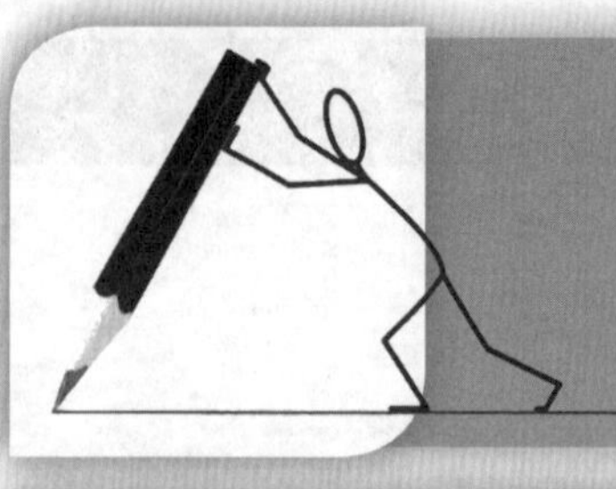

describing

meaning:

giving a detailed account of the properties/qualities/features/parts of something or someone

sentence starters

One of the characteristics of ... is ...

... has a number of distinguishing/special/notable features, including ...

The key features of ... are ...

... looks like ...

An examination of ... reveals ...

... has several distinguishing features, which include ...

The major attribute of ... is ...

... has some distinctive features/characteristics that make it unique.

... comprises/is composed of/consists of/is constructed of ...

Other important aspects are ...

The most significant elements of ... include ...

... has some very distinctive traits, especially ...

... is unlike anything else seen/experienced previously, although ...

A cursory glance reveals several strengths/weaknesses.

Upon examination, it is seen that ...

This shows that ...

The most obvious feature of ... is ...

Most prominent is ...

Other less important features are ...

Important though ... is, it is not the most relevant factor in the description.

key task words

define, describe, identify, state

connectives

additionally	all	along with	also	although
and	apart from	as shown in/by	as well as	besides
composed of	extra	for example	furthermore	however
in addition	included	in contrast	mainly	moreover
not only ... but also	rather than	several	shows/showed	some
such as	too			

useful mathematical language

appeared/ appearance	combined	comprised	described/ description	displayed
examined/ examination	features	happened	impact	revealed
occurred	properties	took place	the following characteristics	visually

example

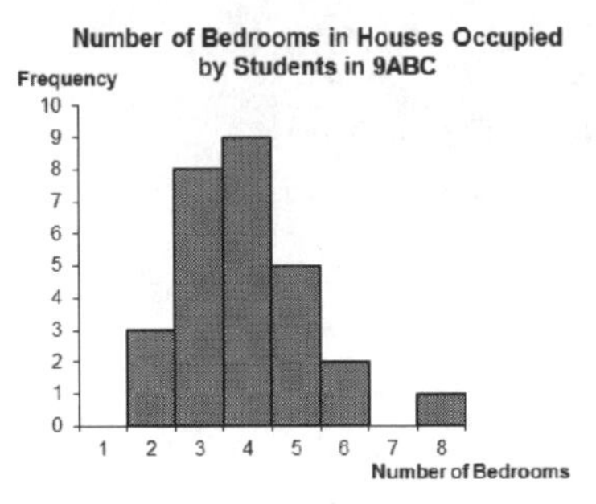

The histogram *shows* the number of bedrooms in houses occupied by the 28 students in class 9ABC. **An examination of the graph reveals** that the mode (most common number of bedrooms) is four (nine students). *All* students live in houses with between two and eight bedrooms.

Other important aspects are that more than half of the students (17 out of 28) lived in houses with three or four bedrooms, *although* eight students lived in houses with five or more bedrooms. One student lived in a house with eight bedrooms. *In contrast,* three students lived in two bedroom houses.

explaining

meaning:

making the reader understand something by giving reasons for '**how**' and/or '**why**' things are as they are

sentence starters

There are several reasons for ...

... has multiple causes, which include ...

... is like this because ...

There are several aspects to the problem to be examined, especially ...

Each part of the problem contributes to ...

Whilst A allows us to identify the cause/s of the problem, B enables us to ...

The factors that contribute to this situation include ...

The main effect of ... is ...

They are most likely to occur when ...

As a result of ..., it will be necessary to ...

The current situation exists because ...

... are usually caused when/by ...

The main reason that ... occurs is ...

There is no apparent reason why ... is like it is; however, ...

The reason for this situation is not clear; however, ...

... works by ...

To put it simply, ... is/are caused by ...

This happens/has happened because ...

If ... were to change, then ... would happen.

The best way to do this is ...

The steps involved are ...

The process usually starts with ...

To achieve this goal, it is necessary to ...

key task words

account for, discuss, examine, explain, interpret, justify

connectives

affect	also	and so	as a result of	because
consequently	due to these factors	followed (that)	from this/these/that	gave rise to
hence	however	if ... then	in order to	may
now that	so	so that	then	this is how
this is why	thus	to ensure that	unless	when ... then

useful mathematical language

accounted (for)	caused	clarified	consequence	created
detailed/details	effect (of)	ensured (that)	explained/ explanation	happened
interpreted/ interpretation	led (to)	outcome	prompted	provided (that)
reason (for)	resulted (in/from)	supported (by)	upshot	

example

A budget is a plan of future income and spending. **The main reason that** households use budgets is *to ensure that* they do not plan to spend more than they earn.

A household budget usually starts with a list of all of the family's income. The most common form of family income is wages and salary, but it may also include interest earned on savings and benefits paid by the Government. The next step is to list everything that the family spends money on. *Because* it is often a very lengthy list, it can be difficult to remember everything. **For this reason**, many families keep detailed records of their spending in the year before. *If* the total spending is greater than the total income, *then* the household will not have enough money to meet all of its obligations. **As a result**, the household may need to cut back its spending plans.

Initially a family budget is put together for a whole year. *However*, it is *then* split up into detailed plans for each month. **This happens because** there may be many expenses in some months, but not very many in others. The budget allows families to plan to save money in the 'good' months *in order to* have enough available to pay the bills in the 'bad' months.

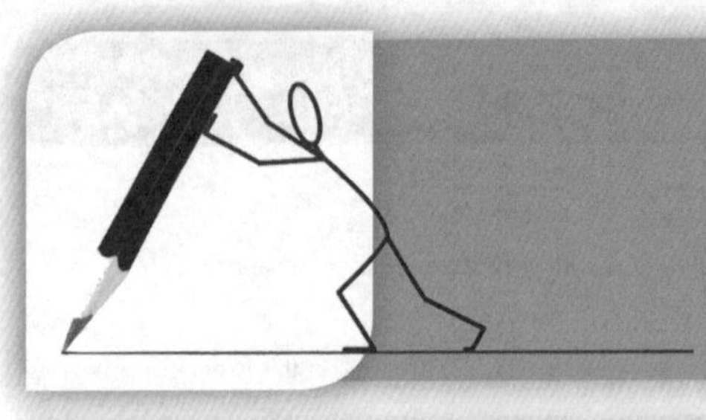

extrapolating

meaning:

using known facts about something as a basis for general statements about a situation or suggesting what is likely to happen in the future

mathematics meaning:

extending a graph to obtain additional values

sentence starters

The graph shows that ...

The general characteristics of ... are ...

The data shown in ...

The relationship between ... and ... is generally ...

From the evidence, it is possible to suggest the following generalisation/s:

As ... increased/decreased, so did ....

There was an inverse relationship between ...

In the vast majority of cases ...,

Whilst there are exceptions to the rule, for the most part ... applies.

There are exceptions to this general trend although, for the most part, ... is relevant/evident/significant.

There are anomalies; however, a general pattern emerges which is ...

The graph was extrapolated by the addition of a trend line/line of best fit.

This trend can be extrapolated beyond the range of the data ...

The extrapolated graph allowed predictions to be made about ...

Extrapolation allows predictions to be made beyond ...

Care must be taken when extrapolating graphs in this way.

There would be a point at which the relationship ceases to apply.

key task words

extrapolate, extend, predict

connectives

allows	assumed (that)	because	clear	for example/instance
however	generally	if	it follows that	may/could be
particular	resulted (in)	showed	so	specifically
suggested	therefore	when	where	whilst

useful mathematical language

beyond the data	extended/extension	extrapolated/ extrapolation	increased/decreased	inverse
limited/limitation	line of best fit	linear/non-linear	mathematical model	plot
predicted/prediction	reasonable	regression line	reliable	scatter graph/ scattergram
scope	theoretical	trend		

example

The data shown in Graph 1 was collected in an experiment in which the extension of a spring was measured when different forces were applied to the end of the spring.

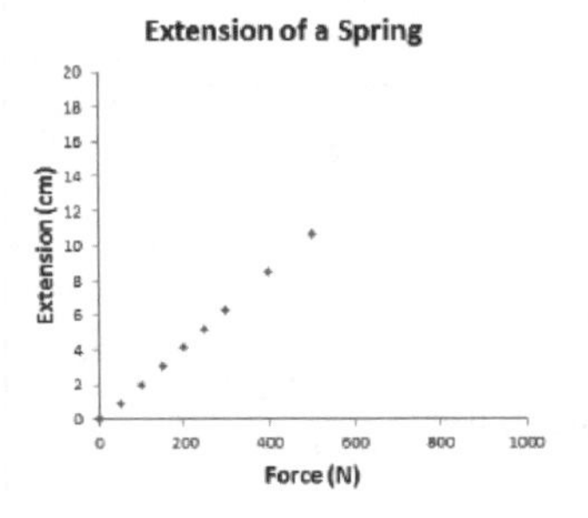

Graph 1

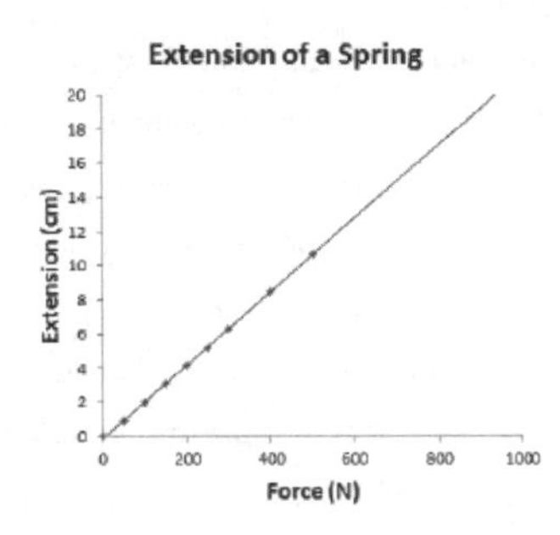

Graph 2

The graph shows that there is a clear linear trend. **As the force increased, so did** the length of the spring. **This trend can be extrapolated beyond the range of the data,** as shown in Graph 2. *Whilst* the largest force applied in the experiment was 500 Newtons, **the extrapolated graph allows predictions to be made about** the extension of the spring when larger forces are applied. *For example,* if a force of 700 Newtons was applied, the spring may be extended by 17 centimetres.

Care must be taken when extrapolating graphs in this way. The further the trend line is extended, the less reliable the predictions become. For example, there would be a point at which the spring would be fully extended. Additional force would not add to the length of the spring. It may be that the spring could break under very large forces.

Extrapolation is a useful technique to **make predictions beyond** the scope of the data. However, care Is needed to ensure that the predictions are reasonable.

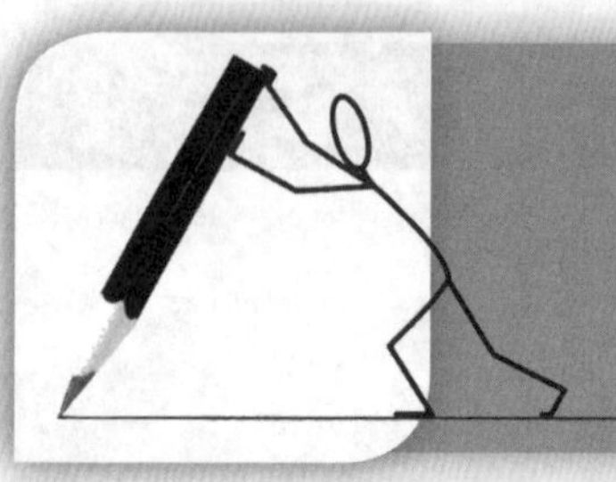

generalising

meaning:

developing a broad statement that seems to be true in most situations or for most people; does not include details such as evidence or examples

mathematics meaning:

using particular examples to develop an equation or mathematical model to describe the overall situation

sentence starters

It can be seen that ...

Inspection of ... reveals that ...,

This suggests that ...

An examination of the data shows that ...

A general statement can be made about ...

As ... increases/decreases, ... increases/decreases

Generally speaking, ... is relevant in this case.

In this instance, the general rule of ... applies.

So, generalising from these particular cases, ...

From the evidence, it is possible to suggest the following general rule:

In a more general way, the ideas can be represented as:

Algebraically, it could be written as ...

... is hypothesised to be ...

The rule/formula can be checked using the other cases ...

The general rule/formula can be applied to ...

This mathematical model can be used to explain ...

key task words

develop, explain, extend, generalise, predict, summarise

connectives

because	however	it follows that	particular	so
specifically	therefore	this suggests (that)	where	

useful mathematical language

abstracted/ abstraction	algebraic/ algebraically	broad/broadly	case	checked
commonly	correctly	extended/extension	formula	generalised
graph/graphically	hypothesised (that)	limitation	mathematical model	overall
pattern	repeatedly	resulted (in)	rule	showed (that)
substituting/ substituted	tested	theoretical	typical	usual/usually
validated	verified	was	weakness	

example

The angle sum of a polygon is the total size of the angles formed inside the polygon. For example, a triangle contains three angles that add to 180°. The table below shows the angle sum measured in various types of polygons.

polygon	number of sides	angle sum
triangle	3	180°
quadrilateral	4	360°
pentagon	5	540°

polygon	number of sides	angle sum
hexagon	6	720°
heptagon	7	900°
octagon	8	1080°

Inspection of the table reveals that as the number of sides in the polygon increases, so does the angle sum. *Specifically,* the angle sum goes up by 180° for each extra side in the polygon. *However*, the angle sum for a triangle is not simply 180° x 3. This *results in* 540°, which is 360° too large. *This suggests that* the rule for the angle sum of a triangle could be 180° multiplied by the number of sides less 360°. **Algebraically, it could be written as** $180n - 360°$, where n is the number of sides in the triangle.
The rule can be checked using the other five types of polygons shown in the table. *Substituting* the number of sides for each polygon into the rule confirms that it does *correctly* predict the angle sum of each polygon. **So, generalising from these particular cases,** the angle sum of a polygon with n sides **is hypothesised to be**

$$A = 180n - 360.$$

This general rule can be used to predict the angle sum of a polygon without the need to measure the angles.

inferring/ interpreting

meaning:

using what is provided to make meaning or arrive at an answer; to uncover the answer even though it is not directly said or stated

sentence starters

It can be inferred that ...

It means that ...

It is unclear what ... means, but a plausible explanation is ...

It is reasonable to assume that ...

The interpretation of ... is supported by ...

This interpretation is supported by the following evidence:

The diagram shows that ...

The data reveal the following trends:

A trend was observed ...

The ... increased/decreased as ... increased/decreased

The relationship between ... and ... was ...

There was a connection between ... and ...

The graph/table/result showed/revealed that ...

An exception to the trend was ...

It is likely that ... because ...

There are several interpretations, including ...

... has several possible interpretations; however, ... is the most likely

key task words

assume, calculate, conclude, determine, extrapolate, infer, interpolate, interpret

connectives

although	as a result	because	clearly	despite
due to	even though	evidently	far from	hence
however	in contrast	in other words	it is clear	it is evident
may mean that	might mean	since	therefore	this is how
this is why	whereas			

useful mathematical language

appeared/ appearance	assumed (that)	caused	concluded (that)	conjectured
consequently	deduced/deduction	identified	implied	indirect
inferred/inference	interpreted/ interpretation	pattern	reason (for)	resulted
revealed	seemed	surmised		

example

Total over 0.05	Approx. over 0.08	State
26 918	14 805	NSW
16 946	9 320	VIC
5 946	3 270	TAS
10 339	5 639	SA
32 532	10 636	WA
32 532	17 893	QLD
111 933	61 563	Total

Source: http://www.britzinoz.com/drinkdriving-statistics-in-australia

The table shows the numbers of people charged with drink-driving in Australia (state by state) in 2011. *Although* Queensland is the third largest state in population size, it has by far the largest number of drink-drivers. **There are several interpretations of this fact**. **It could mean that** Queenslanders are heavier drinkers than their fellow Australians in other states. **Even though not directly stated**, the figures *may mean that* Queensland police are far more diligent than law enforcers in other parts of the country. The figures for all states are alarming *and reveal* a great deal about the effectiveness – or lack thereof – of safety campaigns. *Despite* aggressive campaigns to stop drivers from drinking and driving, *it can be inferred* that an alarming number of motorists repeatedly risk their lives and the lives of others by breaking the law.

logically justifying

meaning:

giving all the logical reasons and/or mathematical arguments that have led to a decision or outcome

sentence starters

To prove that ...

To show/demonstrate/deduce that ...

The logical argument to support the result that ... is as follows:

It can be demonstrated/proved that ...

This argument is supported by ...

The proof of this hypothesis is ...

This result can be justified by the following deductions:

The calculations show that ...

... is justified by the following argument:

... thereby proving the original proposition.

Using the general rule/proposition that ...

Relying on the proposition/rule/theorem that ...

This proposition can be proved by accounting for all possible examples.

... is a sound/valid argument.

Mathematically ...

The mathematical reasoning is ...

Presenting this argument in mathematical notation ...

Considering a simple example first ...

A pattern can be identified that ...

This pattern can be generalised to conclude that ...

Extending this logic/pattern ...

Using similar logic, it can be concluded that ...

There are no valid logical reasons to support this conclusion.

The following counter-example demonstrates that the proposition/hypothesis is incorrect:

The proposition/hypothesis can be disproved by ...

There is a value of x such that

Let x be ....

..., where x is ...

connectives

always	as	as a result	because	for all
for each/ every value of x	gives	hence	however	if ... then
is supported by	it can be shown that	it follows that	let x be	must be
now	results in	show/shows (that)	since	so
therefore	there is	thus	where x =	whilst

useful mathematical language

add/subtract/ multiply/divide	argued/argument	assume	calculate/calculations	confirm/confirms
correct	counter-example	deduce/deduction	demonstrated (that)	demonstrates/ demonstration
disproved by	find	justify/justification	logic/logical/logically	plausible
produces	proposition	proves/proof	reasonable/reasoning	similarly
substantiates	supports	using	validate	verify
working (out)				

key task words

account for, deduce, demonstrate, disprove, justify, logically argue, prove, show, support, verify

example

MATHEMATICAL CALCULATION:

To find the area of a basketball court that is 28 m long and 15 m wide.

Using the general rule that $A = l \times w$,

$A = 28 \times 15$ (*where* $l = 28$ and $w = 15$)

$= 420$

Therefore, the area of the basketball court is 420 m^2.

PROOF:

To prove that the sum of two consecutive whole numbers is always odd.

Let n be any whole number.

Therefore the next whole number after n is $n+1$

Adding these two consecutive numbers gives

$$n + n + 1 = 2n + 1$$

Since n is a whole number, *it follows that* $2n$ is an even number and *hence* $2n+1$ *must be* an odd number.

Thus, the sum of two consecutive whole numbers is *always* odd.

providing evidence

meaning:

referring to sources, illustrations and other evidence about something to support the points that have been made

sentence starters

Analysis of the data suggests ...

The evidence reveals ...

The graph shows that ...

It is clear from the evidence that ...

As shown by the information in ...,

As seen in Diagram A, ...

Table B shows ...

According to the figures in Table A, ...

The point is well supported by the available data.

Table A and B show conflicting information, viz., ...

Compared with the data in Table A, the data in Table B shows ...

... can be supported by the information in Graph A.

This/these argument/s are confirmed by several authors including (author, date), who states that ...

According to (author, date) who states that ...,

(Author 1, date) argues that ... and this is supported by (author 2, date).

Several authors (author 1: date, author 2: date, author 3: date) are in agreement about ...

The evidence collected allows the following observations to be made:

key task words

demonstrate, exemplify, expound, extrapolate, identify, refer

connectives

according to	also	as well as	clearly	closely
for example	for instance	for this reason	furthermore	in addition to
in detail	indeed	is confirmed (by)	it is clear that	means
moreover	such as	thorough	viz	

useful mathematical language

based on	case in point	contradicted	evident/evidently	illustrated
indicated (by)	indicators	refuted	revealed	showed
suggested	substantiated	supported		

example

Graph A shows that the unemployment rate among Indigenous Australians is three times higher than the national average, **and Graph B shows** that their average income is about half. **It is clear from the evidence** that they have a higher infant mortality rate (three times the national average); an alarming suicide rate, (six times higher than for non-Indigenous Australians); and an adult life expectancy at least 20 years lower than the average for Australians generally. This *is confirmed by* the fact that 53% of Indigenous men and 41% of Indigenous women die before the age of 50 compared with 13% and 7% non-Indigenous Australians. *In addition to* these factors, there are other indicators of disadvantage. Indigenous Australians living in remote areas of Northern Australia experience extremely high death rates that are three to four times the national average.

It is clear that Indigenous Australians continue to be severely disadvantaged. **According to Smith (2008)**, rates of cardiovascular disease, respiratory illness, diabetes, injuries and infectious diseases among this group are much higher than for non-Indigenous Australians. *As well as* having a higher unemployment rate, the chance of imprisonment is also higher for Indigenous Australians. *In addition*, they are more likely to be homeless or living in overcrowded housing. The situation for Indigenous Australians is dire, a finding **well supported by the available evidence.**

N.B. Graphs are not included

introducing

meaning:

starting a text by describing what it is about, including the aims/objectives, why it matters, key definitions, important issues, and what comes next

sentence starters

It has been suggested that ...

A common view is that ...

There are two main views about ...

It was hypothesised that ...

The background to this issue was ...

This study/investigation/experiment aimed to ...

The objective of the study/investigation/experiment was ...

The method used to explore this issue was ...

The issue was important because ...

The investigation/experiment is interesting/significant/useful because ...

The theory relevant to this investigation/experiment is ...

Research suggests that ...

(Author 1, date) stated ... and this directly supported the findings of (author 2, date), who noted ...

Before continuing, it is necessary to explain the meaning of some key terms.

The important concepts in this study were ...

This report contains ... sections, as follows:

An outline of this report is as follows:

key task words

describe, foreshadow, introduce, outline

connectives

any	between	for example	for instance	if any
important	included	meant (that)/ meaning	on the one hand	on the other hand
significant	such as	the reason for	was defined as	was made up of

useful mathematical language

aimed/aim	comprised	connection	defined/definition	examined
explored	goal	important/ importance	included	intended/intention
investigated	is in the third section	issues	mattered/matters	objective
purpose	relationship to	significant/ significance	the final section	the next section
this section				

example

Every school offers students the opportunity to engage in co-curricular activities *such as* sport, music, drama, hobbies, debating or community service. **There are two main views** about the impact of these activities on academic results. *On the one hand,* **it has been suggested that,** students who choose to become involved in many school activities achieve better academic results than those who see school only as a place for study. *On the other hand*, some argue that co-curricular activities take up time that should be spent studying and therefore reduce the students' grades.

This study aimed to explore *any* statistical *connection between* involvement in co-curricular activities and academic results. **The method used to explore this issue was** a survey of a group of students in Year 9 in Somewhere High School.

Involvement in co-curricular activities *was defined as* voluntary participation since the start of Year 9, in school-based activities in the students' own time. Academic results were based on the grades in each of the eight subjects included in the most recent school report.

This report contains five sections *including* this introduction. *The next section* describes the methods used to conduct the survey. The analysis of the data *is in the third section*, just before a section discussing the results. *The final section* summarises and concludes the report and makes recommendations for the school, based on the findings of the survey.

sequencing

meaning:

Involves putting things in the order in which they happened or will happen. This style of writing in a mathematical report can be called a *method* or *procedure*.

sentence starters

Before beginning, it was necessary to ...

Initially, ... occurred.

The first event that happened was ...

During the first stage of ... the following event/events happened/occurred:

This was followed by ...

Once ... was completed/put in place/decided, then ...

The second/next step/process involved ...

The next logical step was to ...

In the next stage of the (process) ...

There was a defined sequence of events that was adhered to.

It mattered/did not matter if ... occurred first.

Following ..., ... commenced.

The key events in the order in which they happened are listed below:

It was important that these steps occurred in the prescribed order.

The final part of the process ...

On completion, ...

key task words

list, order, outline, sequence, summarise

connectives

after	also	and then (use sparingly)	as a final point	as before
as soon as	as well as	at the outset	at the same time	each/every time
finally	first, second, third, etc.	formerly	in addition to	in summary
in the beginning	initially	lastly	meanwhile	moreover
now that	on top of	once	previously	prior to
so far	subsequently	ultimately	until now	

useful mathematical language

began/beginning	commenced	concluded	datum (singular)/ data (plural)	equipment
experiment	following/followed by	investigation	measured	planned
preceded	process	progressed/ progression	order	recorded
sequence	stage	started (with)	step	succession
survey	table/tabulated	tallied/tally		

example

Before beginning the survey, **it was necessary to** do some planning. *First*, the sample size was determined to be one quarter of the students in Year 9. **The second process was** the selection of a sample by drawing students' names randomly out of a box containing the names of all Year 9 students. **This was followed by** the design of a questionnaire, used to ask for details about the student (name and class), the number of co-curricular activities that they were involved in, and their Semester 1 grades in each subject (see Appendix A). *Finally*, a tally sheet was designed to record the students' answers (copy at Appendix B).

The next stage of the survey was data collection. Students were interviewed in the first week of Term 4 during their mathematics lessons, by the six students conducting this project. The students' answers were recorded in the tally sheet. **The final part of the process** was the analysis of the data, described in the next section.

[Appendices A and B are not included in the example.]

analysing

meaning:

using statistical techniques to summarise, compare or infer something. This style of writing in a mathematical report can be called *data* or *analysis*.

sentence starters

Some of the data collected in this experiment were inaccurate because ...

Some observations/data were excluded because ...

Initially, the data were converted from ... to ...

This allowed/enabled ...

It was assumed that ...

The mean/median/mode of ... was calculated/determined to be ...

The information was detailed/summarised/ presented in ...

The data were recorded in ...

... was/were measured/recorded

The information/result was tabulated/graphed ...

The main similarities/differences in the data were ...

The data reveal the following trends:

A trend was observed ...

The ... increased as ... increased/decreased.

The ... declined as ... increased/decreased.

The relationship between ... and ... was ...

The data revealed a strong connection between ...

This means that ...

The graph/table/result showed/revealed that ...

There is an inverse relationship between ... and ...

... is inversely proportional to ...

key task words

analyse, calculate, compare, convert, describe, extrapolate, graph, infer, interpolate, interpret, summarise, tabulate

connectives

although	as well as	because	differ/different	for example
for instance	however	in addition	in other words	in particular
more/less than	on the other hand	respectively	similar to/ different from	

useful mathematical language

accompanied (by)	analysed/analysis	associated (with)	calculated/ calculation	changed/ increased/ decreased/ etc
compared	consisted of	converted	datum (singular)/ data (plural)	estimated
extrapolated	facts	evaluated	graphed/graph/ graphically	graphically
inferred/inference	information	interpreted/ interpretation	interpolated	measured
recorded	showed that	tabulated/table	which was/ allowed/gave	

(see also page 41 for alternatives to 'showed that')

example

Initially, the data were converted from the raw form in which they were collected into numerical values. The number of co-curricular activities **was recorded** as a score ranging from 0 to 6. In each subject, academic results **were measured by** converting the A to E grades to a five point scale from 5 to 1, *respectively*. Each student had one co-curricular score and eight academic scores.

The mean of the academic scores was calculated for each student to give an overall score called the grade point average, ranging from 1 (low) to 5 (high). The co-curricular scores and grade point average **were summarised in a table and a graph** (shown in Appendix C).

For students with a co-curricular score of 0, the **mean** grade point average **was calculated.** This process was repeated for the groups of students with co-curricular values of 1, 2, 3, and so on. **The results were tabulated and graphed** (also shown in Appendix C) and then examined for patterns.

The data revealed a strong connection between grade point average and the co-curricular score. *In particular* the grade point average **improved as** the co-curricular score **increased** from 0 to 3, and then levelled off. *In other words*, up to three additional co-curricular activities seemed to have been accompanied by increases in the grade point average. *However*, four or more co-curricular activities did not appear to be associated with any further improvements in the grade point average.

[Appendix C is not included in the example.]

discussing

meaning:

to consider both sides of an issue about something, without necessarily coming to a conclusion

mathematics meaning:

to consider possible explanations of mathematical results or analysis. This style of writing in a mathematical report can be called *results* or *discussion*

sentence starters

The examination of the data revealed that ...

The evidence suggested that:

In the usual course of events ...,

It is likely that ... because ...

A number of general statements could be made about ...

In the vast majority of cases ...,

Whilst there are/were exceptions to the rule, for the most part ... applied.

Although there was some variation, a general pattern that emerged was ...

Whilst there were some exceptions, for the most part, ... was relevant/evident/significant.

This interpretation is supported by the following evidence:

A strength of this approach is ...

Based on this survey, ...

There are several possible explanations for this, including ...

Whist the survey/experiment/investigation did not collect information to explain this, it may have occurred because ...

... has several possible interpretations; however, ... is the most likely.

Generally speaking, ... was relevant to this case.

To summarise, then, the following trends were apparent:

It means that ...

It can logically be inferred that ...

It is reasonable to assume that/apply this finding to ...

... was more usual than ...

This finding could also be applied to ...

key task words

argue, discuss, explain, extend, generalise, outline

connectives

as likely as not	because	essentially	for the most part	generally
in common	in general	in most/many cases	in the main	likely
mainly	many	more likely than	more often than not	most
most often	mostly	not	often	on average
on the whole	on the one hand	on the other hand	primarily	reasonably
regularly	resulted (in)/results	showed (that)	suggested (that)	since
typically	usually	was/were		

useful mathematical language

assumed/ assumption	broad/broadly	case	characteristically	checked
commonly	considered	converse	correctly	extended/extension
found/finding	generalised	graph/graphically	hypothesised (that)/ hypothesis	limitation
mathematical model	overall	outcome	pattern	repeatedly
strength	theoretical	typical	usual/usually	validated
verified	weakness			

example

The examination of the data revealed that more co-curricular activities *were usually* linked to higher academic results. There are two possible explanations. *On the one hand*, it could be that those students with enthusiasm, personal organisation and commitment chose to be involved in co-curricular activities. Those qualities can also lead to academic success. *On the other hand*, it may be that students learnt skills in co-curricular activities, such as team work and higher order thinking, which also helped them to improve their academic results. **Based on this survey**, it is not clear whether one or both explanations are correct. A wider survey would be needed.

Whilst there are exceptions to the rule, *for the most part*, higher academic results did not continue indefinitely. There seemed to be little extra benefit, in terms of academic results, from participation in four or more co-curricular activities.

It is reasonable to apply this finding to the whole of Year 9 in Somewhere High School as the sample was chosen randomly from Year 9 students in the school. However, a wider survey, involving students in other year levels and/or other schools would be needed before these results could be extended to those other groups.

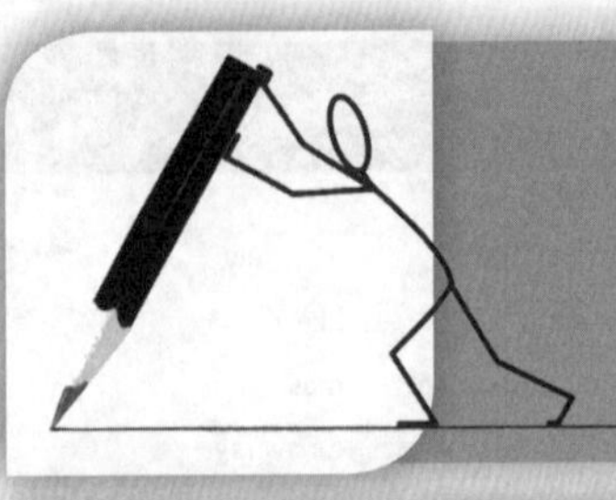

recommending

meaning:

suggesting a course of action for consideration by others; providing reasons (usually the findings of an investigation) in favour of the suggestion

sentence starters

It is recommended that ...

It would appear reasonable to conclude that ..., and therefore recommend ...

In spite of ..., the best solution is ...

Based on the findings, the following recommendations are suggested:

There are too many problems associated with ...; therefore, ... is recommended.

The recommendations, after looking at all the evidence, are ...

It is, therefore, advisable to propose that ...

The following recommendations, listed in order of priority, are put forward for consideration:

Based on the analysis of the situation, the following recommendations have emerged:

The following changes are recommended:

For the future, it is recommended that ...

In light of the problem of ..., a suitable solution should include ...

To achieve its goals, the organisation needs to ...

The sensible option is to ...

The following recommendation should be enacted as soon as possible:

A list of the proposed recommendations with a summary of the reasons for each follows:

In the light of all available data, the proposed recommendation is ...

There are a couple of recommendations from which to choose.

If the following recommendation/s is/are implemented, then the problem should be solved.

There are no guarantees that these recommendations will work; however ...

key task words

construct, devise, prepare, present, propose, recommend, suggest

connectives

accordingly	all things considered	as a result	because	consequently
could/should/would	finally	general	hence	however
if ... then	in future	in summary	in the end	in the final instance
is desirable	is expected to	is favoured	is recommended	moreover
on balance	should	then	therefore	these include
thus				

useful mathematical language

achieve/achievement	act/action	advise/advice	advocate	benefit
conclude/conclusion	improve/improvement	propose/proposal	recommend/recommendation	suggest/suggestion
summarise				

example

If co-curricular activities help students to learn skills that improve their academic results, *then* the school *should* aim to include more students in co-curricular activities. *Accordingly*, **the following recommendations** for the school **are proposed**:

1. Advise students and parents of the benefits of co-curricular activities.
2. Increase the range of co-curricular activities offered.
3. Explain that the time taken up by involvement in co-curricular activities does not appear to affect academic results.

There are no guarantees that these recommendations will improve academic results; *however*, co-curricular activities have many other benefits for students.

concluding

meaning:

drawing together the main ideas of something and restating them in a succinct way, often as a decision; a conclusion may involve making recommendations for the future. In the case of a report of an experiment or a survey, the conclusion may also explain whether the aims were met.

sentence starters

A conclusion can be drawn from ...

Thus, to conclude/in conclusion ...

It can be concluded that ...

To sum up, ...

In summary, ...

It would appear reasonable to conclude that ... and, therefore, recommend ...

An examination of all the data allows the following conclusions to be drawn: ...

The analysis supports the conclusion that ...

It is not clear ...

The most valid conclusion is that ...

Consequently, it would seem better to ... than ...

Considering all the options, it would seem better to ...

In summary, ...

An examination of the evidence allows the following summary to be made: ...

A strength of this conclusion/approach is ...

The conclusions of this study/investigation are limited to ...

These conclusions could also be applied/extended to ...

Given that the aim of this study was to ..., this report shows that this objective was partially/fully met.

The aims of the study could not be fully met because ...

The following unexpected outcomes occurred:

A further study could extend this work by ...

Further studies would be required to ...

key task words

conclude, examine, investigate, summarise

connectives

accordingly	although	because	conclusively	consequently
decisive/decisively	finally	if	in brief	in closing
inconclusively	inevitable/ inevitably	much doubt	no doubt	now
on balance	on condition that	overwhelming/ overwhelmingly	so	some doubt
then	these/ which include	thus	whether ... or	

useful mathematical language

applied (to)/ application	assumed (that)	clarified/clear	concluded/ conclusion	consequence
effect (of assumptions)	expected	extended (to)/ extension	final point	found/finding
goals	impact	importance/ important	limited (to)/ limitation	outcome
results	significance/ significant	summed (up)/ summarised/ summary	unexpected	was

example

The aim of this study was to explore any statistical link between involvement in co-curricular activities and academic results amongst Year 9 students in Somewhere High School. **In summary,** it found that there was evidence of a positive link between participation in co-curricular activities and school grades. **It is not clear** *whether* this link was because students with higher grades get involved in co-curricular activities more often, *or* because involvement in co-curricular activities developed skills that led to higher academic results.

The conclusions of this study are limited to the group surveyed (the Year 9 students in Somewhere High School). However, if these students are typical of those in other Year levels and/or other schools, **the conclusions could be extended to** these groups. **Further studies would be required to** confirm this.

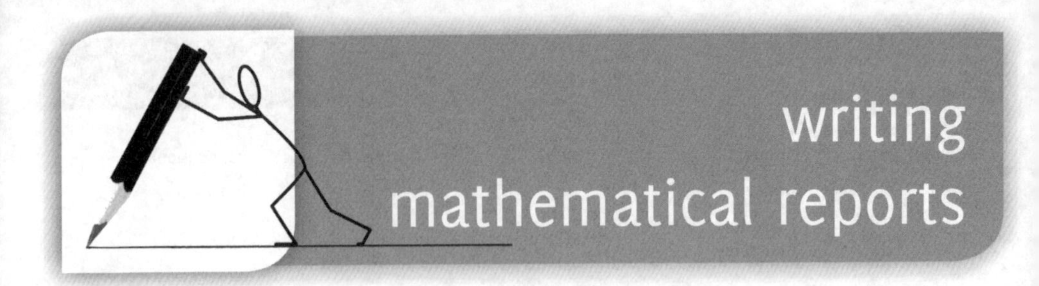

writing mathematical reports

A mathematical report requires several different writing skills. Each section of the report should be drafted separately, with the different sections put together into one document just before the report is finalised. The graphic organiser on page 39 assumes that the essential sections of the report are: introduction, method/procedure, results/analysis, discussion and conclusion. Other sections, such as abstract, recommendations, and appendices, are optional.

The organiser can help in planning the report. Each section links back to the writing skills detailed earlier in this book. The layout of the graphic organiser suits the order in which the sections are usually drafted: the procedure, results and discussion sections should be drafted first. Then the conclusion and recommendations (if any) should be developed. The introduction should be drafted after the other sections have been written. The last section to be written is the abstract (if required). Any research required for the project might be included in different parts of the report: introduction, procedure, results or discussion.

The main sections of a report (introduction, procedure, results, discussion, recommendations and conclusions) should be as short as possible. All details, such as calculations, tables, graphs, and diagrams, should be placed in appendices. However, the information in each appendix must be summarised in the main body of the report, so that the reader can understand what the report is saying without having to read the appendices. Each appendix must be mentioned somewhere in the main body of the report so that the reader knows that further supporting information is available.

Headings (and possibly sub-headings) should be used to signpost the different sections of a report. It is usual to number the sections.

GRAPHIC ORGANISER FOR PLANNING A REPORT

Title:

What is the name of the report?

Research:

What information is needed? Collect information for the bibliography during the research.

Procedure: (sequencing, see page 28)

- *Consider numbering these steps.*
- *Where/when were the data collected?*
- *Put copies of survey forms in an appendix.*

Results: (analysing, see page 30)

- *Put raw data in an appendix.*
- *Use tables and graphs to summarise the results.*

Discussion: (discussing, see page 32)

Conclusion(s): (concluding, see page 36)

Recommendation(s): (recommending, see page 34)

Appendices:

For detailed information. As the report is prepared, make notes of the details to be attached. Use these notes as a checklist for collating the final report.

Introduction: (introducing, see page 26)

*Write this section **after** the conclusion and any recommendations, but put it **before** the procedure.*

Abstract:

*Write this section **last**, but put it **before** the introduction.*

Other Details:

✓ Does the report contain your name, class, teacher's name?
✓ Are any of the following needed: task sheet (put this on top), title page, table of contents?
✓ Has the report been proof read?
✓ Are all the pages numbered and collated in the correct order? Are they all stapled together?

The shaded sections may be left out, depending on the requirements of the task.

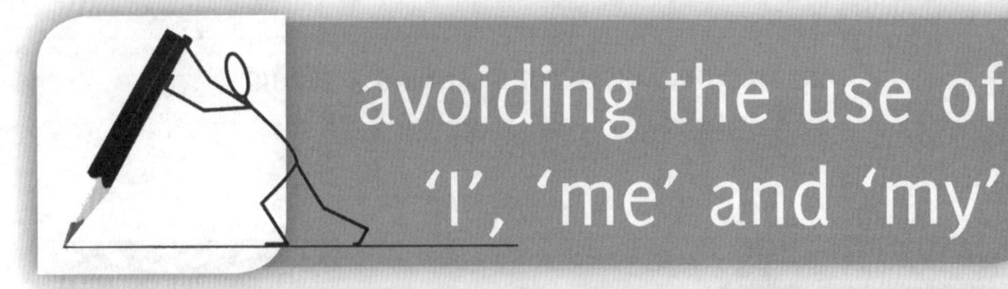

explanation:

In most forms of formal, academic writing, the use of the **first person** (I, me or my) is generally discouraged. More authority is given to the work if it is written in the **third person.** In addition, the use of the **passive voice** (i.e., the 'doer' is removed) is also a prominent feature of more technical texts.

ways of saying

It was found that ...

It could be suggested that ...

This is/can be illustrated by ...

It is seen through ...

This is evident when ...

Upon examination, it becomes apparent that ...

The facts indicate that ...

This is exemplified by ...

This illustrates that ...

This shows that ...

Therefore, it can be stated that ...

Clearly, this becomes apparent when ...

With some exceptions, sources generally agree that ...

... clearly points out that ...

This is most obvious when ...

It can, therefore, be observed that ...

There is evidence to support both opinions on this topic.

Author 1 (date) is in total agreement/disagreement with author 2 (date) when ...

Most notable exceptions to this rule are ...

Observations reveal that ...

substitutes for 'showed that'

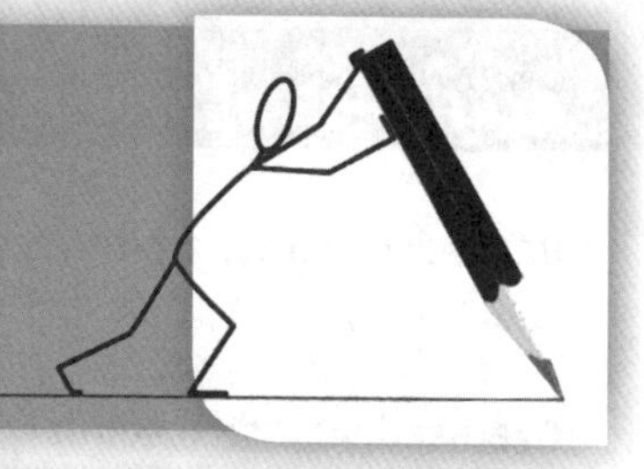

explanation:

It is common practice in mathematical writing to refer to sources of information such as data, diagrams, tables, graphs, and research by others. Relying on phrases such as, The graph showed that ..., the table showed that..., the information showed that ... etc., can be a very tedious and unsophisticated form of writing.

ways of saying

... affirmed that

... claimed that

... confirmed that

... considered that

... demonstrated that

... depicted the

... determined to be

... displayed that

... established that

... exemplified the

... exhibited the

... explained that

... exposed the

... found that

... gave the result that

... illustrated the idea that

... implied that

... indicated that

... was an example of

... led to the conclusion that

... manifested the

... presented the

... proved that

... provided evidence of

... recorded the

... reflected the

... represented the

... resulted in

... revealed that

... suggested that

... validated the result that

... according to ...

... as could be seen in ...,

... was consistent with ...

... it could be seen in ... that

examples of words to avoid in mathematical writing

Imprecise words: a bit; a few; a little; a lot; about; around; heaps; highly; loads; lots; many; masses; moment; roughly; several; some

Overused words: absolutely (unless part of technical term); actually; amazing; awesome; fine; good/bad; got; interesting; like; literally; major (unless part of a technical term); nice; quite; really; said; very; well

Foreign words and phrases: inter alia; prima facie; faux pas; fait accompli; a priori

Personal pronouns: I; me; you; she; her; he; him; it; we; us; you; they; them; myself; yourself; himself; herself; itself; ourselves; yourselves; themselves; my; your; his; her; its; mine; yours; his; hers; our; your; their; ours; yours; theirs

Contractions: aren't; can't; couldn't; didn't; doesn't; don't; hadn't; hasn't; haven't; he'd; he'll; he's; I'd; I'll; I'm; I've; isn't; it's; let's; mightn't; mustn't; shan't; she'd; she'll; she's; shouldn't; that's; there's; they'd; they'll; they're; they've; we'd; we're; we've; weren't; what'll; what're; what's; what've; where's; who's; who'll' who're; who's; who've; won't; wouldn't; you'd; you'll; you're; you've

Double negatives: I can't get no satisfaction

Prolixity (unnecessary words): actually; after due consideration; as a matter of fact; as far as ... is concerned; basically; basis; on the basis that; by and large; cannot help but; in (under) the circumstances; comparatively; conditions; currently, presently; cut back; doubtless, undoubtedly; due to; effectively; (due to) the fact that; goes without saying; in the final analysis; hitherto; if and when; literally; marginally; may well be; needless to say; not necessarily; obviously; overly; period of; personally; in point of fact; point (moment) in time; in the process of; put an end to; the reason is because; reason why; relatively; resulting from, as a result of; in spite of; there is no alternative but; to all intents and purposes; to the tune of; utilise; very; as to whether; well and truly; whether or not

examples of words to avoid in mathematical writing

Clichés: a far cry from; abject failure; acid test; across the board; ample opportunity; as a matter of fact; back to basics; back to the drawing board; ballpark figure; be all and end all; benefit of the doubt; big picture; bottom line; by and large; by the book; cut both ways; cutting edge; despite my best efforts; dos and don'ts; easier said than done; ground breaking; high hopes; ins and outs; last but not least; law of averages; lion's share; moment of truth; part and parcel; proof of the pudding; proving ground; reinvent the wheel; saving grace; stumbling block; take pains; test of time; the exception proves the rule; the root of the matter; tip of the iceberg; to all intents and purposes; to and fro; too numerous to mention; tried and tested; user-friendly; wait and see; winning streak

Tautologies (words that mean the same thing): absolutely essential; added bonus; advance planning; advance warning; all-time record; ask the question; bits and pieces; blend together; brief moment; cancel out; circle around; close proximity; combine together; could possibly; drop down; each and every; estimated at about; final outcome; first and foremost; first priority; full to capacity; gather together; join together; lag behind; lift up; like ilk; might possibly; minute detail; necessary requirement; never before; new innovation; none at all; open up; past experience; period of time; plan ahead; present time; puzzling problem; raise up; reason why; reiterate again; revert back; short summary; small in size; still persists; sum total; ten in number; the vast majority; true facts; twelve noon/midnight; ultimate goal

conventions for showing mathematical working

Mathematical symbols are a precise form of shorthand. Symbols (such as the + symbol) that have a specific mathematical meaning are reserved for mathematical use. Using them correctly is a skill that takes time to develop.

Numbers: There are nine numerals: 0, 1, 2, ... , 9. A particular numeral can have different meanings, depending on how it is used (consider the meaning of the 3 in: 3; 3154; 0.3; ⅓, 4^3). Numbers with more than four digits are spaced in groups of three (without commas) to aid understanding (e.g., 123 456 789). Numbers written in a column should be right aligned (if whole numbers) or aligned under the decimal points.

Units of measurement (e.g., mm, kg) do not form part of a calculation. They should be introduced at the end of the calculation.

Pronumerals (including variables and unknowns) are usually represented by upper or lower case letters from the English or Greek alphabets. Letters from the English alphabet are commonly typed in script form (Times New Roman italics works well) to avoid confusion with other uses of these letters. Letters such as upper case 'I' and 'O' and lower case 'o' are not generally used to avoid confusion with the numerals 0 and 1. A new pronumeral should always be defined, describing what it is being used to represent. An exception would be a constant such as π. Once a variable has been assigned a meaning, it cannot be re-used in the same context with a different meaning.

Calculations involving several steps: If a process requires more than one step, each step should go on a separate line. In each line, the = sign should be aligned. Blocks of calculations are usually indented so that they are towards the middle of the line.

conventions for showing mathematical working

Equations: If an equation is important, it should be placed in the middle of its own line. The most commonly used, and misused, symbol is =. The = symbol means that the things on either side are the same value, just written a different way. The common misuse of = is to mean 'the next step is ...'. For example, when asked to compute (3 + 5) x 2, some people will write (incorrectly): "3 + 5 = 8 x 2 = 16". The problem can easily be seen if the middle step is removed, resulting in the clearly incorrect "3 + 5 = 16". This calculation should be split over two lines, with the first line showing 3 + 5 = 8 and the second line 8 x 2 = 16. If the left hand sides of consecutive equations on different lines are the same, it can be omitted on the second and subsequent lines. It follows that a line can begin with =. However, a line must not end with an equal sign (or an inequality sign).

Explanations: Key words that explain the reasoning are essential. Examples include *if, so, where, whilst, let, since, because, therefore, for each, for every* (or the mathematical symbols used to represent these words). For a lengthy or difficult process where the reader might not readily follow each step, a phrase or sentence may be needed to explain what is occurring. In such cases, the words should be written on a line of their own. The solution of any problem must end with a concluding sentence (including any units of measurement).

Word Processing: It is almost impossible to type complex mathematical processes in the correct form using a standard word processor and keyboard. Software is available to assist in typing mathematical notation correctly (such as Equation in Microsoft Word and MathType), but it can take a while to become familiar with its use. Many computer applications use mathematical symbols in forms that can be typed easily on a standard computer keyboard (eg: *, /, ^). However, these symbols are **not** an acceptable alternative to the correct mathematical symbols.

degrees of intensity (modality)

MODE	LOW →							HIGH
probability	impossible/ impossibly	improbable/ improbably	unlikely	possible/possibly	likely/ in all likelihood	probable/probably	sure/surely	certain/certainly
frequency	never	seldom	occasional/ occasionally	sometimes	often	usual/usually	regularly/ in most cases	always
certainty	never	scarce/scarcely	perhaps/ in some cases	might/could	as likely as not	inevitable/inevitably	undoubted/ undoubtedly	define/definitely
extent	never	scarce/scarcely	limited	partly	general/generally	mainly	almost	complete/ completely
confidence	suspect	unreasonable/ unreasonably	doubtful/doubtfully	moderate/ moderately	reasonable/ reasonably	plausible/plausibly	undeniable/ undeniably	irrefutable/ irrefutably
emphasis	quite	really	conceivable/ conceivably	very	sure/surely	definite/definitely	incredible/incredibly	absolute/absolutely
importance	desirable/desirably	prefer/preferably	require/required	necessary	important/ importantly	unquestionable/ unquestionably	essential/essentially	vital/vitally
intensity	scarce/scarcely	slight/slightly	mild/midly	intermittent/ intermittently	moderate/ moderately	typical/typically	unrelenting/ unrelentingly	extreme/extremely

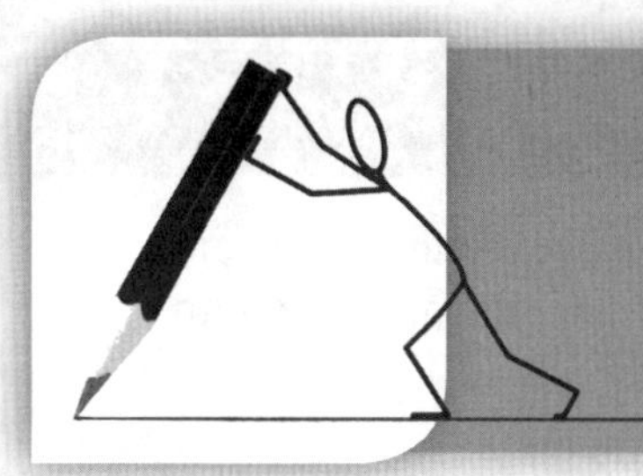

key task word glossary

abstract	to create a general idea about something rather than one relating to a particular object, person or situation
account for (maths)	to report on; to try every possibility
account for	to give reasons for something and report on those reasons
analyse (maths)	to use statistical methods to summarise, compare or infer something
analyse	to examine the parts of something in detail and discuss or interpret the relationship of the parts to each other and to the whole; may involve description, comparison, explanation, interpretation and critical comment
appraise	to consider something or someone carefully and form an opinion about them
argue/ persuade	to present one or both sides of an argument and use persuasive techniques to convince others that your opinion about something is the correct one
arrange	to place things into a particular position, often with a degree of order or precision
assess	to make a judgment about something based on its value or worth (may include quality, outcomes, results or size)
assume	to accept that something is true without necessarily confirming it or checking its validity
break down	to make a list of the separate parts of something
calculate (maths)	to obtain a result from given facts, data or other numeric information about something
calculate	to ascertain or determine something from facts, figures or information
categorise	to divide things or people into sets or say to which set they belong based on common criteria

key task word glossary

classify	to group things with similarities in the same classes or categories; to defend the inclusion of similar things into these categories
comment on	to present your opinion about something
compare	to examine two or more things or people and note the ways in which they are similar **and** different
conclude	to draw together the main ideas of something and restate them in a succinct way, often as a decision; a conclusion may involve making recommendations for the future
consider (maths)	to ensure that your response refers to the particular information you have been given about something
consider	to give opinions in relation to the information you have been given about something or someone
construct	to make, build or put together items or arguments about something
contrast	to examine two or more things and focus on the differences
criticise	to make judgments about something or someone, giving details to support your views
decide	to choose something or someone based on a consideration of other possibilities
debate	to examine both sides of an issue about something and come to a conclusion - or leave the reader/listener the opportunity to come to a conclusion
deduce	to reach a conclusion about something based on evidence that is known to be true
defend	to argue in support of something
define (maths)	to give the meaning or precise description of the concept
define	to show, describe or state clearly what something is and what its limits are
demonstrate	to show something by example
describe	to give a detailed account of the properties/qualities/ features or parts of something or someone
devise	to have an idea for something and design and plan it

key task word glossary

differentiate (maths)	to find a derivative
differentiate	to recognise or show the differences between one thing or person and another
discriminate	to recognise that two things or people are different
distinguish	to draw attention to, and make note of, the distinct differences between things or people
discuss (maths)	to consider possible explanations of mathematical results or analysis
discuss	to consider both sides of an issue about something, without necessarily coming to a conclusion
elaborate	to give more information or detail about something or someone
evaluate (maths)	to find the exact value of something
evaluate	to give a considered judgement about the value or worth of something or someone, and support it with evidence
examine	to look at something carefully, often for reasons "**how**" or "**why**" something may have happened
exemplify	to give more information or details about something
explain	to make the reader understand something by giving reasons for both "**how**" and "**why**" things are as they are
expound	to present a clear and convincing argument for a definite and detailed opinion about something
extend	to include or affect other people or things
extract	to obtain information from a larger amount or source of information
extrapolate (maths)	to extend a graph to obtain additional values
extrapolate	to use known facts about something as a basis for general statements about a situation or suggesting what is likely to happen in the future
generalise (maths)	to use particular examples of something to develop an equation or mathematical model to describe the overall situation

key task word glossary

generalise	to develop a broad statement that seems to be true in most situations or for most people; this does not include details such as evidence or examples
identify	to notice or discover the existence or presence of something or someone
illustrate	to use examples of something to give more detail to information or more weight to an argument
indicate	to point out something from available information
infer/interpret	to use what is provided to make meaning or arrive at an answer; to uncover the answer even though it is not directly said or stated
investigate	to examine the reasons for something
justify (maths)	to give all the logical reasons and/or mathematical arguments that have led to a decision
justify	to show or prove that a decision, action or idea about something is reasonable or necessary by giving sound, plausible and logical reasons for it; to answer the question **"why"**
list	to arrange related items in order, usually under one another
order	to arrange things in a logical way
outline	to give all the main ideas about something without the details
paraphrase	to restate what someone has said or written in a slightly different way from the way it was first stated; the meaning is retained
predict	to suggest what might happen based on the available information
prepare	to gather what you need to make ready for something that is going to happen
present	to put forward something for consideration
propose	to put forward something (for example, a plan, an idea, a point of view, an argument, a suggestion)
prove (maths)	to produce a logical mathematical argument that shows the truth of a statement for all values or situations

key task word glossary

prove	to support something with facts and figures
quote	to repeat words exactly as they appear in the material
recommend	to suggest a course of action for consideration by others; provide reasons (usually the findings of the investigation) in favour of the suggestion
refer	to use material in your answer without necessarily directly quoting from the stimulus material or information
review	to go back over earlier points about something, often with a view to what went wrong or could be improved
sequence	to put things in the order in which things are arranged, actions are carried out, or events happen
sketch	to give the main ideas briefly about something or to create a sketch or drawing that shows the essential features; detail or accuracy is not required
solve	to find an answer or solution to a problem
suggest	to put forward or propose an idea or plan about something for someone to think about
support	to use a fact to support a statement or theory about something
summarise	to give a short account of something with the main points but not the details
synthesise	to put together various elements (from several places or sources) to make a whole; the reassembled material is often original
trace	to show how events/arguments progress and develop
value	to establish the worth of something or someone
verify (maths)	to test the truth of something
verify	to back up a particular result and prove something

teacher reference: the nature of mathematical writing

Mathematical writing is the prose that accompanies or supports mathematical arguments and investigations. Mathematics teachers must explicitly teach this style of writing and not assume that students have learnt it elsewhere, or that it is transferrable from other contexts.

Mathematical writing requires conciseness that is achieved by ensuring that every word 'adds value' (see examples to avoid on page 42). The appropriate use of mathematical terms, symbols, standard mathematical abbreviations, tables and visual images allows the writer to convey information whilst keeping word usage to a minimum. A writing style that uses short sentences, headings and sub-headings will also add to conciseness and clarity.

Readability of mathematical texts.

Reading mathematical texts can be challenging for many students. This is because the reading level of a mathematical text is usually higher that other types of texts. There are two main reasons for this. First, readability is influenced by sentence length. Mathematical writing often needs to use strings of words to convey a single idea (e.g., the extension of the spring), which contribute to longer sentences. Further, writing in passive voice and the third person usually increases the length of a sentence. Second, readability is influenced by the number of syllables in a word. Mathematical vocabulary uses polysyllabic words more frequently than everyday English (e.g., mathematics, denominator, pronumeral, quadrilateral), thus increasing reading levels.

Reading levels in mathematics texts, as measured by standard indexes of readability (such as the Flesch-Kincaid Index used in Microsoft Word), can sometimes appear to be high, although the text is not complex. One reason for this is that the reading level is increased by the inclusion of symbols. For example, the readability, as assessed by the Flesch-Kincaid Index, of a passage of 230 words including eight numbers written as symbols was reduced by a whole grade level simply by removing the numbers. Secondly, writing in point form can result in what appears to be very long sentences, although broken up into several points. Whilst the use of point form makes the text easier to understand, it increases the measured reading level. Standard readability indexes were not designed for use in mathematical texts and, in consequence, are not necessarily a reliable indicator of reading complexity.

my useful words and phrases

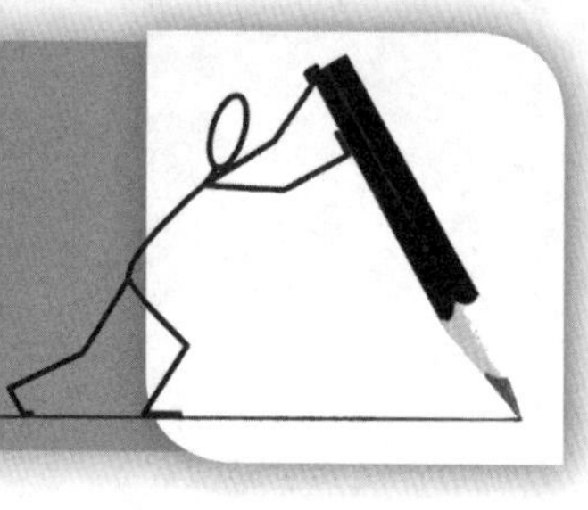

my useful words and phrases

my useful words and phrases

my useful words and phrases

my useful words and phrases

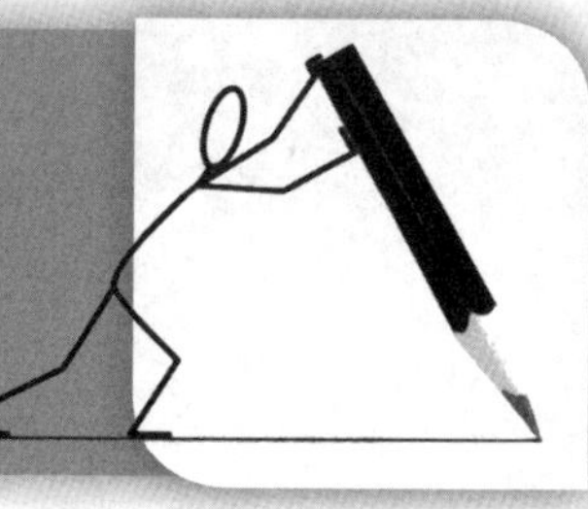

about the authors

Patricia Hipwell M.Ed., B.Sc. Econ. (Hons), Grad. Dip. of Literacy Ed., P.G.C.E. is an independent literacy consultant for her own company, **logonliteracy**. She delivers literacy professional development to teachers in Australia, and works predominantly in Queensland schools. Patricia has specialised in assisting all teachers to be literacy teachers, especially high school subject specialists who often struggle with what it means to be a content area teacher and a literacy teacher. Assessment has been an area of interest for many years and much of Patricia's work enables teachers to create assessment that is 'do-able'. This is important because students often have very little idea of what they are required to do and rely heavily on parents/caregivers to assist them.

Lyn Carter M.Ed., Dip.Ed., B.Ec., J.P., operating through her business **Count on Numeracy**, is an independent consultant, providing professional development to teachers of numeracy and mathematics throughout Australia. She is completing her doctoral studies in NAPLAN numeracy testing. Lyn has several areas of expertise: preparation of students for standardised testing, especially in numeracy; assisting all teachers to become teachers of numeracy; and secondary mathematics. One of her particular interests is the literacy aspects of mathematics. On the one hand, experience and research show that the majority of student errors in mathematics problems arise from difficulties in interpreting the problem. On the other hand, the ability to express mathematical ideas in written form is critical to being a successful mathematician and an important aspect of the interface between mathematics and numeracy.

In addition, Patricia and Lyn are part of the educational consulting team at ITC Publications.

Patricia and Lyn have developed a number of resources to assist students' literacy and numeracy development. Both consultants are available (as a cross-curricular team or individually) to provide professional development to teachers in their areas of expertise and to support the use of their recommended resources, including this one.

For further information, contact:

Patricia Hipwell
Mobile: 0429 727 313
Email: pat.hipwell@gmail.com

Lyn Carter
Mobile: 0402 077 958
Email: countonnumeracy@bigpond.com

itc

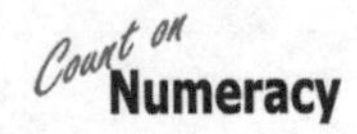